THE ATMOSPHERE

Our Changing Climate
THE ATMOSPHERE

Contents

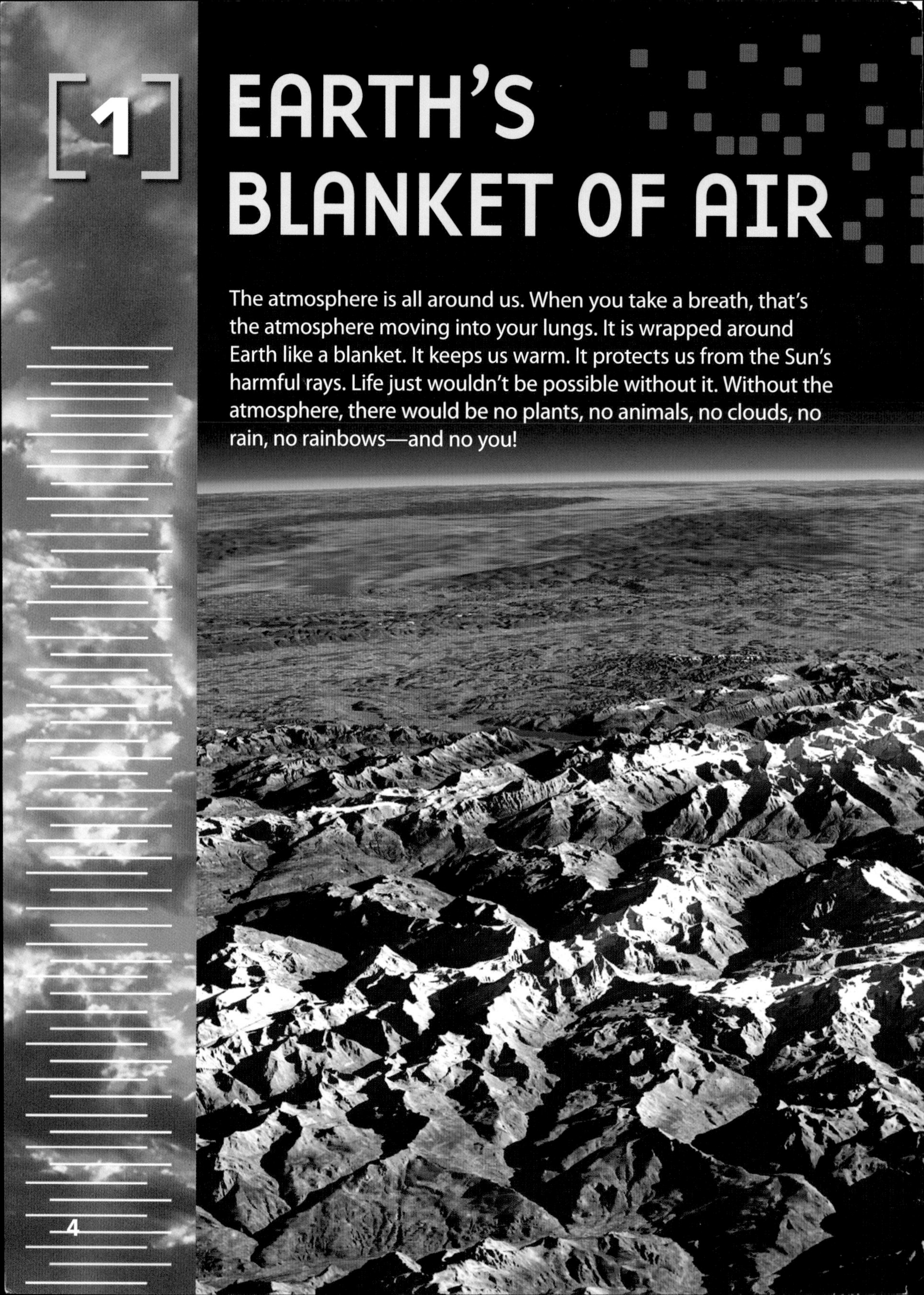

[1] EARTH'S BLANKET OF AIR

The atmosphere is all around us. When you take a breath, that's the atmosphere moving into your lungs. It is wrapped around Earth like a blanket. It keeps us warm. It protects us from the Sun's harmful rays. Life just wouldn't be possible without it. Without the atmosphere, there would be no plants, no animals, no clouds, no rain, no rainbows—and no you!

For the last two centuries, humans have been changing the air. We've been adding gases to the atmosphere that are making our planet warmer. This is changing Earth's climate. And it's affecting not only our air, but also our oceans, our ecosystems—and us.

THE AIR OVERHEAD

Other planets have atmospheres too, but Earth's is different. Our air is a special mixture of gases that make it possible for people, animals, and plants to live here. Without the atmosphere, our Sun would scorch all the plants and animals, and temperatures would plunge below freezing at night.

Layer Cake

The atmosphere is a bit like a giant, invisible layer cake—one that starts on the ground and reaches up into space. The lowest layer is called the troposphere. This is where birds, airplanes, and Frisbees fly. It's also home to clouds, rain, snow, and hurricanes—the world's weather. About 90 percent of our air is in the troposphere, and it's the layer most affected as the climate changes.

The troposphere is 8 kilometers (5 miles) thick above the poles and 16 kilometers (10 miles) thick above the equator.

SPF 1,000

Where the troposphere ends, the next layer, called the stratosphere, begins. It goes up to about 50 kilometers (30 miles), and holds most of the rest of our air. Think all the action's in the troposphere? Think again. The stratosphere is where the ozone layer is. The ozone layer protects all of us on the ground from the Sun's harmful ultraviolet (UV) rays. It's our planet's sunscreen.

4 U 2 Do

So Thin!

How big is the atmosphere compared to Earth? Wrap a basketball or a soccer ball in a recyled plastic bag and knot it at the top. If the ball is Earth, that layer of plastic is the atmosphere. Pretty thin for something so important.

How Do They Know?

Up, Up, and Away!

In the 19th century, people floated high in hot-air balloons to study the atmosphere. As they rose higher and higher, they found that it got cooler and cooler. The temperature of the troposphere decreases slowly as you go up. Planning a balloon ride? Better take along a warm coat.

Ionosphere

Stratosphere

Troposphere

Breathing Thin Air

The temperature isn't the only thing that changes as you go up. The amount of air around you changes, too. The higher you go, the fewer air molecules there are. That's why many people who climb Mount Everest use oxygen tanks to help them breathe.

[**Atop Mount Everest, there's one-third the oxygen available at sea level.**]

The Pressure's On

Air pressure also decreases with altitude, mostly because the number of air molecules decreases. Air pressure is the pressure created by air molecules as they zing around and crash into each other and everything else around them. Each collision gives just a tiny push, but there are so many of them that the collisions create a measurable pressure.

It's a Gas!

So what's our atmosphere made of? Mostly nitrogen and oxygen. Together, they make up about 99 percent of our air. If you trapped some air in a jar and could count the molecules, you'd find that 78 out of every 100 were nitrogen molecules and 21 out of 100 were oxygen.

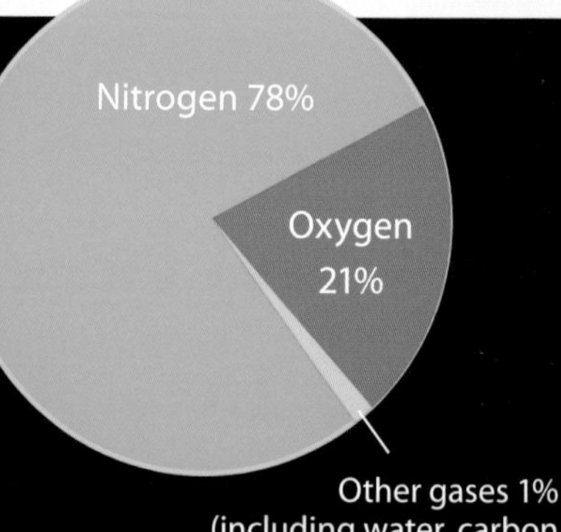

Not Without a Trace

If about 99 percent of all the molecules in our air are either nitrogen or oxygen, do we even care what makes up the other 1 percent? We certainly do! That 1 percent includes carbon dioxide, water vapor, methane, and ozone. These gases make up only a tiny portion of the atmosphere, but they help determine Earth's temperature and its weather.

Don't Breathe Yet

Three billion years ago, microscopic single-celled creatures that lived in the sea were the only living things on Earth. And there was no oxygen in our planet's atmosphere. None. Not exactly paradise. Then some of those primitive microbes evolved the ability to harness the Sun's energy to turn carbon dioxide and water into sugar for food. As part of this process, called photosynthesis, they released oxygen. Oxygen molecules began bubbling out of the water one by one and collecting in the air. Okay, you can breathe now!

Great Invention

Photosynthesis turned out to be a brilliant way to use solar power to create food—and it really caught on! Over the next two billion years, Earth's atmosphere filled with oxygen. Today plants on land and phytoplankton in the ocean carry out photosynthesis. Those early microbes should have applied for a patent.

4 U 2 Do

Edible Atmosphere

What do the atmosphere and trail mix have in common? Mix up your own model of the air. Open your kitchen cupboards and find three different-colored "molecules"—maybe cereal, sunflower seeds, and raisins. Anything about that size will work. Count out 100 pieces, dividing them into three piles based on their percentages in the air—one pile for nitrogen, one for oxygen, and one for the trace gases. How many do you have in each pile? Now mix them all up. That's the atmosphere—a trail mix of gases. Chow down.

Hint: Put your favorite "molecule" into the nitrogen pile.

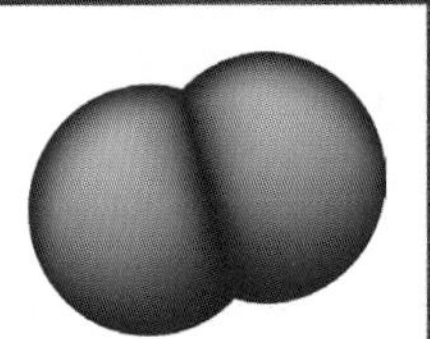

O_2
Oxygen

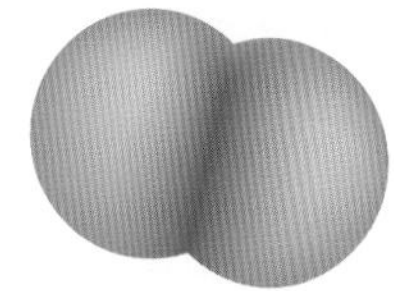

N_2
Nitrogen

Check out your answers on page 40.

[3]

EARTH'S CLIMATE

So what is climate, anyway? Is it just a fancy word for weather? No. The climate is related to the weather, but it's not the same. Weather is what you see when you look out the window. Climate in your hometown is the average weather you can expect where you live. But you can also talk about the climate of a country, a continent, or the whole planet.

What's the Big Idea?

Cloudy, With a Chance of Change

Earth is getting warmer. That means climates around the world are changing. They're not all changing at the same rate or in the same way—but they're all changing. And that's affecting everything on the planet in one way or another.

Ch-ch-ch-changes

Sagebrush, orchids, and ivy. Inchworms, penguins, and salmon. All have adapted to live in particular conditions and particular climates. Now it's getting warmer almost everywhere. Snow and ice are melting. Dry parts of the world are getting drier. Storms are becoming more intense. And people, plants, and animals everywhere are having to move or adapt to the changes.

Royal penguins live together in large colonies.

How Do They Know?

Climate Detectives

Scientists are playing modern-day Sherlock Holmes as they gather evidence about the climate using instruments on the ground, in the air, in the oceans, and even in space. The clues are all around us. Weather stations measure air temperature, wind speed, and rainfall. Ocean buoys surf the waves, and underwater robots dive deep to measure water temperature, salinity, and ocean currents. Sophisticated satellites measure rainfall and sea surface temperatures—even over remote parts of the planet. Not so elementary!

Power Up

That big yellow ball in the sky, the Sun, powers our climate. The Sun is constantly emitting energy in all directions—and fortunately, a small part of it falls on Earth. Sunlight provides the light and heat that we depend on to live.

Starts with the Sun

The Sun powers our climate. Sound simple? Not a chance. Sunlight interacts with the air, land, and water in different and complicated ways. Those interactions drive the circulation of air and water around the planet, the movement of heat from the tropics to the poles, and weather patterns everywhere.

Reflect on This

Not all the sunlight that comes our way actually warms the planet. Huh? Well, some is bounced right back out to space, reflected by ice on the ground or by certain clouds into the sky. The ice sheets and glaciers covering the North Pole, Antarctica, Greenland, and other frigid parts of the planet act like mirrors—another way that ice keeps us cool!

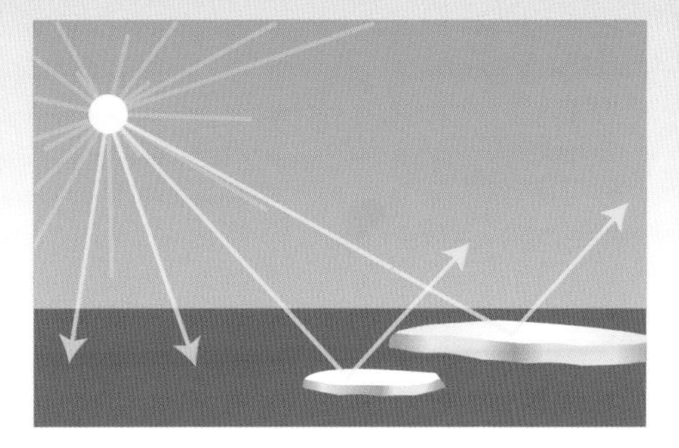

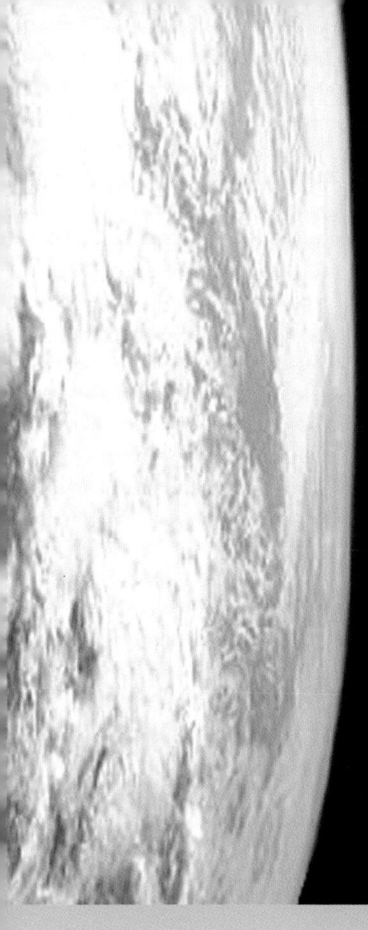

Cloudy Vision

Earth is a cloudy place. But how cloudy? To know how much of the sky is cluttered with clouds, you need to know the cloud cover over the whole Pacific Ocean, the huge Sahara Desert, and the frigid continent of Antarctica. It's hard to find enough volunteers to observe the weather in all those places! Ah, but it's the perfect job for satellites in space. They tell us that, on average, about 60 percent of Earth is covered by clouds, and that clouds reflect about 20 percent of the Sun's radiation before it hits Earth. But these are rough estimates—satellites confuse clouds and ice. Still, satellites are the best eyes we have for a global view of these wispy puffs of water vapor.

Inside the Greenhouse

The sunlight that does strike our oceans and land is absorbed at the surface and warms our planet. The warm surface then tries to cool off by radiating the heat back toward space. If this heat could make it out through the atmosphere as easily as the sunlight makes it in, our planet would be much colder than it is. But not so fast! A few gases in the atmosphere—the greenhouse gases—absorb some of the heat before it escapes into space. They trap the heat and make our planet warmer than it otherwise would be. Yes, this is the greenhouse effect (right).

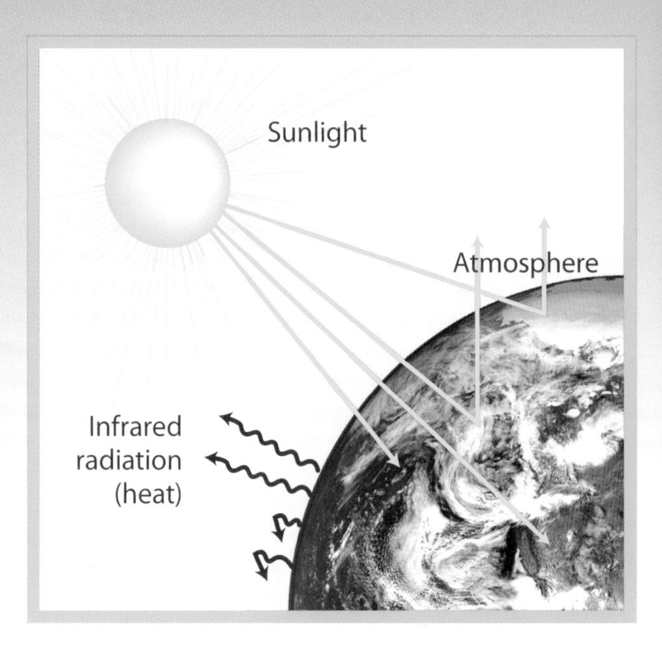

The Other 1 Percent

Not all gases are greenhouse gases. In fact, almost all of our air is made of gases that are *not*! If oxygen and nitrogen made up 100 percent of our atmosphere, instead of 99 percent, there would be no greenhouse effect on our planet. You might think that would be a good thing. Think again. Remember the trace gases in your trail mix?

More Zambonis

The most important greenhouse gases are water vapor, carbon dioxide, and methane. They are nothing new. They were wafting in Earth's air long before there were people on the planet. And though they're only a tiny percentage of our air, those few molecules provide a greenhouse effect that warmed Earth long before humans began to change the atmosphere. In fact, if there were no water vapor or carbon dioxide in our air, Earth would be about 33°C (59°F) colder than it is! Our planet would be one giant ice rink.

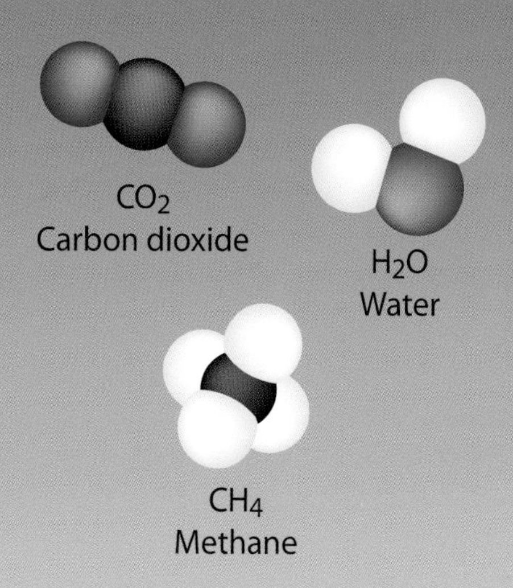

"Earth to Venus"

How big a difference can the greenhouse effect make? Just visit Venus—but be sure to bring a cooler. Venus (right) is the hottest place in our solar system—hotter than Mercury, even though Mercury is the closest planet to the Sun! And it's all because of a huge greenhouse effect. Without its greenhouse effect, Venus would be hot, but livable—about like Phoenix on a very hot summer day. But Venus has a thick atmosphere made almost entirely of carbon dioxide. The resulting greenhouse effect makes the temperature well over 427°C (800°F) all year round—almost 390°C (700°F) hotter than it would be without the carbon dioxide. Sizzling.

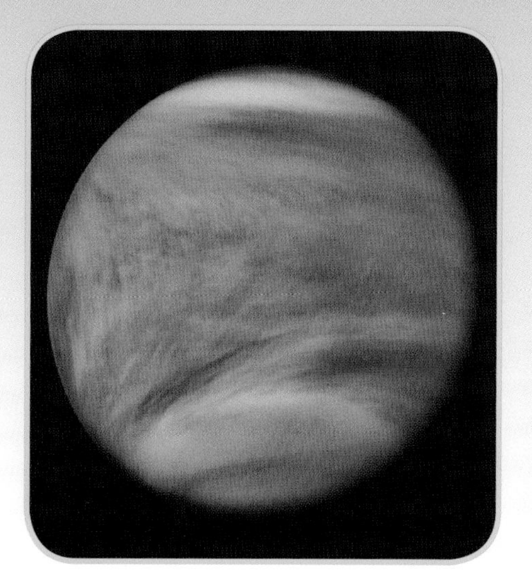

CO_2—FRIEND AND FOE

If the greenhouse gases in our atmosphere are what keep us from becoming one big ice sheet, why are we so concerned about adding more of them? It's because those gases that we're sending into the air are piling up and causing even more warming. And that's affecting our whole planet.

Back to the Future

Climate change is nothing new. Over Earth's long history, there have been cooler times, like during the ice age that happened 15,000 years ago, when ice sheets covered the land where Chicago is today. Very long ago, there were warmer times, like the tropical dinosaur days, which ended 65 million years ago. In the past, these climate changes were usually triggered by natural shifts in the Earth-Sun orbit or variations in the amount of sunlight reaching parts of Earth. That's not the case today.

Ready for Ripley's

This time, climate change is different. Humans are the cause. How did we do *that*? We've changed the atmosphere . . . much faster than it's ever been changed before. Many of the things we do—driving cars, flying in planes, making things in factories—add greenhouse gases to the atmosphere. And we're adding lots of them. Believe it.

The Good Old Days

Before around 1750, people didn't have much effect on the atmosphere. For one thing, there weren't that many of us on the planet. For another thing, there were no industries, no smokestacks, no power plants, no electricity, and no cars. Since then, well, things have changed.

What's the Big Idea?

Room for One More?

The number of people on Earth has skyrocketed. Before 1750, there were fewer than 1 billion people sharing our planet. Now there are more than 7 billion of us. That number goes up by nearly 80 million people every year—that's 25 more people on the planet every 10 seconds!

Burning coal provides half the electricity to keep the lights on in cities like Chicago.

Every Little Bit

All those people are adding greenhouse gases to our air. We've added carbon dioxide by burning fossil fuels like gasoline and oil. We travel by car, ship, and plane. We light our cities, heat our homes, fuel our factories, and power our televisions and computers. Have you turned on a light, microwaved some popcorn, or played a video game lately? Then you, too, have been adding your share.

How Do They Know?

Aloha, CO_2

Before 1958, no one knew how much carbon dioxide was in the atmosphere. That year, a young scientist named Charles Keeling set up a monitoring station near the top of Mauna Loa, the largest volcano in Hawaii, to find out. He picked that location so that his instrument would capture fresh air off the ocean, far away from the traffic and smokestacks of a big city. He measured the amount of carbon dioxide in the air above Mauna Loa continuously for many years. His measurements were used to create one of the most famous graphs in science.

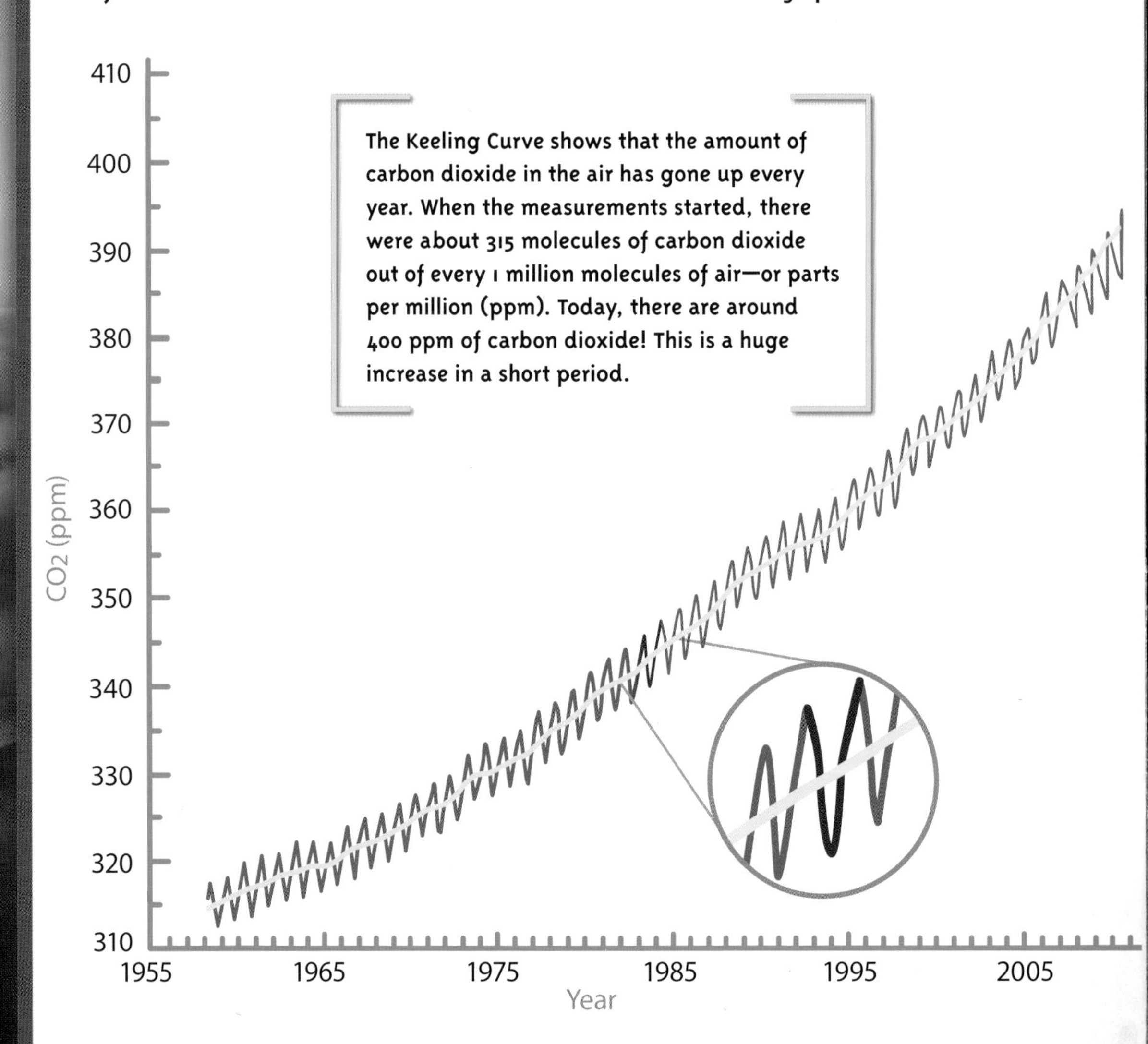

But What Are Those Squiggles?

Charles Keeling figured out this, too. The clue was that the amount of carbon dioxide went down in the spring and summer, then back up in the fall and winter—and that it did the same thing every year. Then he figured out the reason—plants! The graph goes down in the spring, when plants in the Northern Hemisphere grow and bloom and suck in carbon dioxide as part of photosynthesis. It goes back up in the fall, when leaves drop from the trees and many plants go dormant. The squiggles show Earth "breathing." But Earth's breathing is the opposite of ours—it inhales carbon dioxide and exhales oxygen.

How Bad Is It?

Scientists know from records of our climate in the past that the amount of carbon dioxide in the atmosphere and the temperature are linked. When one goes up, the other goes up; when one goes down, the other goes down. Today, the carbon dioxide level is around 400 parts per million (ppm). That's higher than it's been in over 650,000 (yes, 650,000!) years.

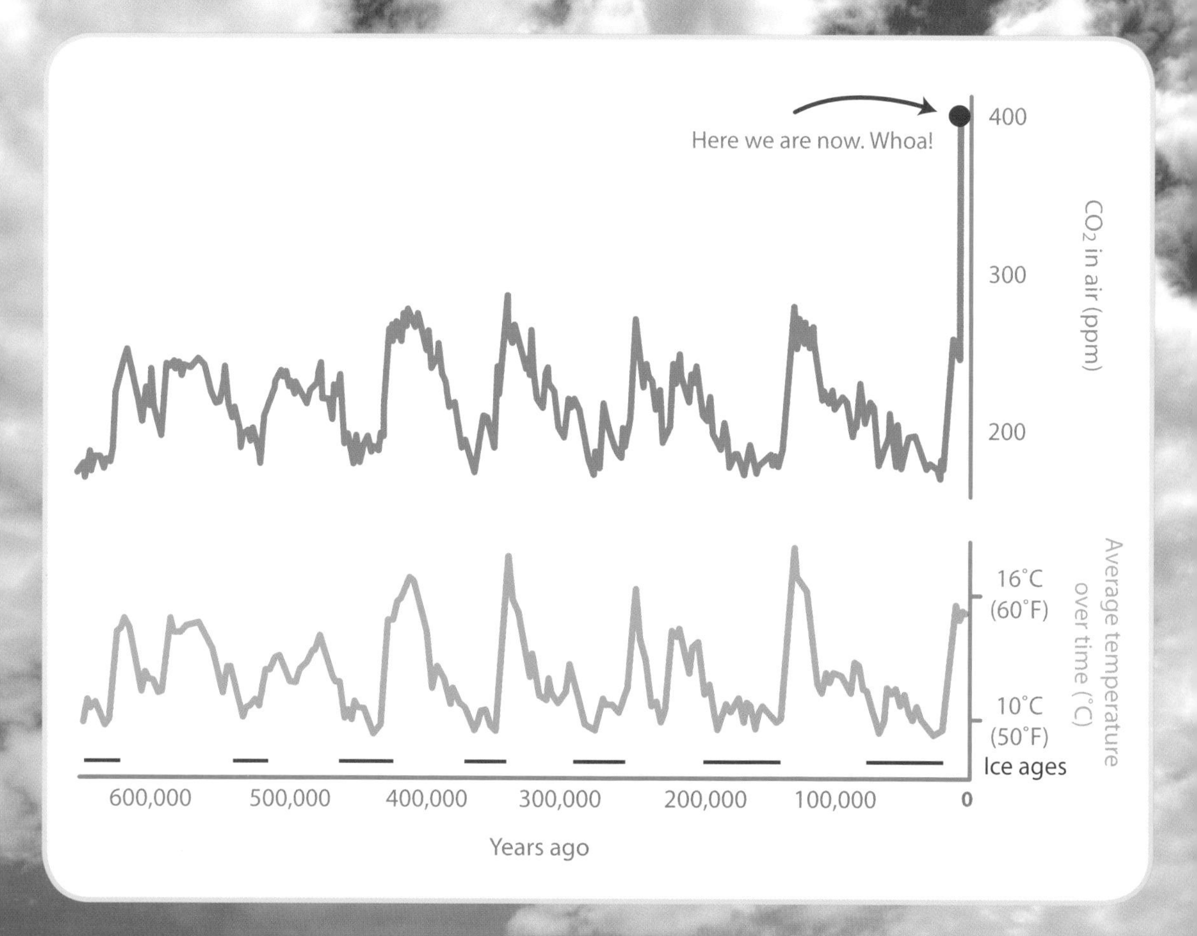

How Do They Know?

Blowing Bubbles

How do scientists know what the air was like thousands of years ago? They look down—into the ice. In Antarctica, it's so cold that the snow never melts. It just builds up year after year. As one layer falls on another, small pockets of air get trapped. The snow slowly gets packed into ice, and the pockets of air become tiny bubbles of air. Scientists are able to drill deep down into ancient ice and pull out long cores of it (top). When they find bubbles trapped in ice (bottom), they can analyze the air inside. They know how old that air is because they know how far down it was in the ice. They've found air that's hundreds of thousands of years old!

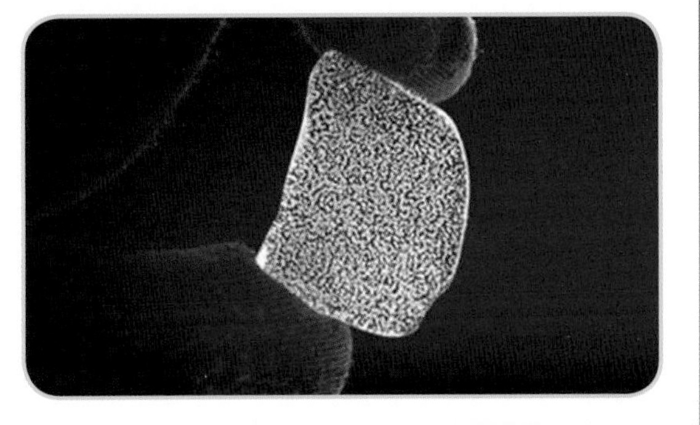

When Cows Belch

Did you know that eating grass makes cows and sheep burp? Each burp is a cloud of methane. Rice paddies, swamps, landfills, and sewage treatment plants belch methane, too. There is twice as much methane in the air as there was in 1750! Why? Because there are eight times more humans on the planet, and we raise way more animals and crops and produce megatons more waste than we did back then.

When Good Gases Go Bad

Seemed like a good idea at the time. Fertilizers were created to increase crop yield on farms, but it turns out they give off nitrous oxide. Yep, it's a greenhouse gas. CFCs, or chlorofluorocarbon gases, were a breakthrough—they were early coolants for refrigerators and air conditioners. But CFCs have been implicated in damage to the ozone layer, and they're also powerful greenhouse gases. Who knew?

A Sky Full of Water

The atmosphere helps move water around our planet—transporting it through the sky and delivering it to thirsty lakes, rivers, gardens, forests, and farmlands—giving us fresh water to drink and rain for our plants to grow.

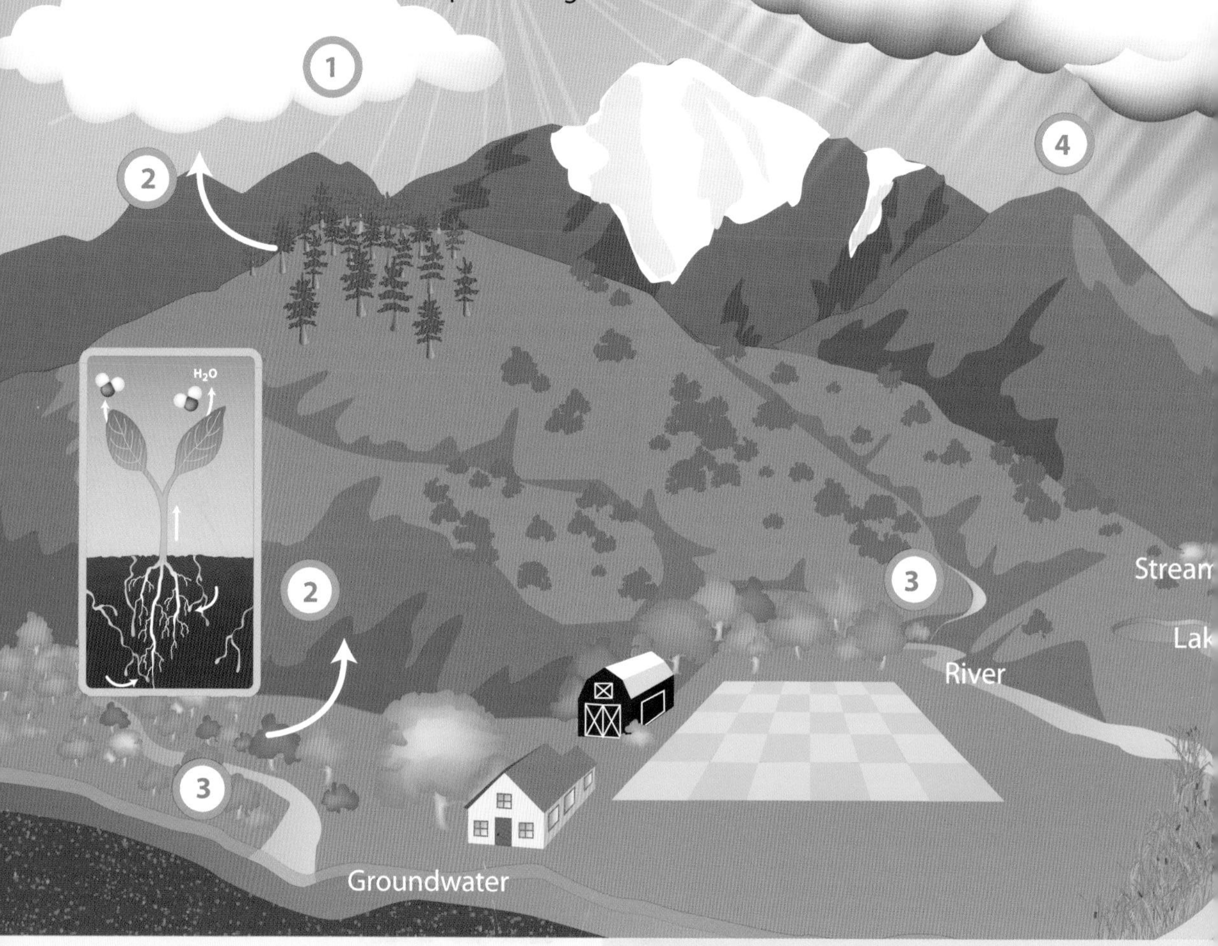

H_2O 007

A water molecule is like a resourceful international spy—always on the move, zipping around the world, traveling on the wind one day, in an ocean current the next, and constantly changing identities. Ice, today. Water, tomorrow. Water vapor, the day after. What's next—secret codes and fake passports?

Floating on Air

When water on land or in the ocean is warmed, some of the molecules escape into the atmosphere as a gas. They evaporate. Not surprisingly, most evaporation is from the oceans. But don't worry—they have plenty of water to spare.

The Water Cycle

1. Condensation
2. Transpiration
3. Water flows downhill and returns to rivers, lakes, and oceans
4. Precipitation
5. Evaporation

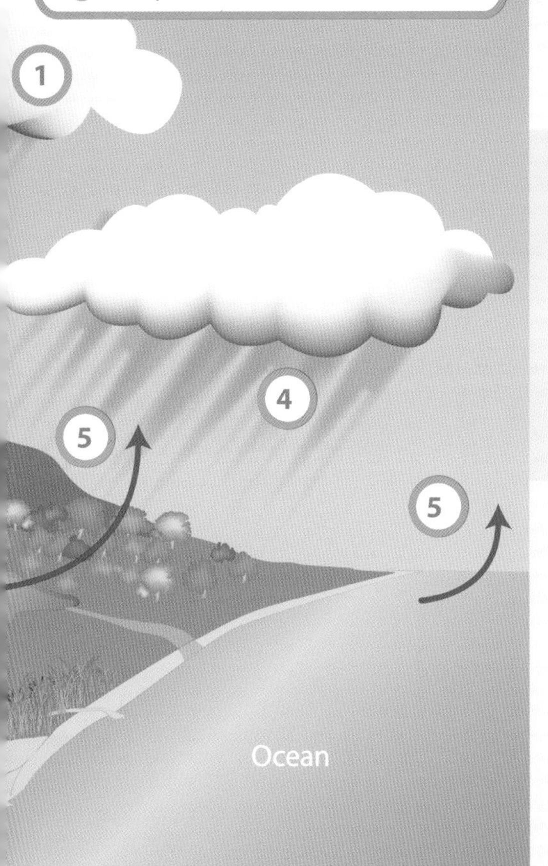

Sweaty Palms

Did you know that plants *sweat*?! Plants around the world are constantly losing water through tiny holes in their leaves. It's called transpiration, and it accounts for about 10 percent of the water released into the atmosphere each year. A single oak tree releases 151,417 liters (40,000 gallons) of water into the air in a year.

Where's That Umbrella?

When water vapor cools down, it morphs back into liquid and condenses into small droplets. When billions of tiny droplets congregate, they form a cloud. Winds can blow clouds over great distances, redistributing water around the planet. It takes lots and lots of those tiny droplets to form a single raindrop, but eventually the water falls back to Earth as rain or snow.

Marathon Cycling

The water that rained out your picnic has already been cycling around the world for *quite* a few years—billions of years, in fact. Hey, you may have sipped water once slurped up by a pterodactyl. *Pterrifying*!

Change Is in the Air

The water cycle never stops—but it *can* be profoundly altered. As global temperatures rise, the water cycle is speeding up. Warmer air means more evaporation. Soil and plants dry out more quickly, so areas that don't get much rain get even drier. More ocean water gets sucked into the air. Clouds will dump a lot of that extra water on places that our weather patterns already saturate with rain.

4 U 2 Do

Evaporation in Action

Take two glass jars and fill them halfway with water. Put a lid on one of them, and put them both in the sun for a couple of hours. Predict what you will see when you come back.

Check out your answers on page 40.

Measuring Up

Compared to daily shifts in weather, climate change is subtle and hard to measure. In fact, it took scientists years to even be sure it was real. But it is. In the last century, Earth's climate has warmed up about 0.8°C (1.5°F). That may not sound like much, but it's the fastest our planet's global average temperature has changed in a thousand years (below). Scientists predict that Earth will keep warming up—and even faster.

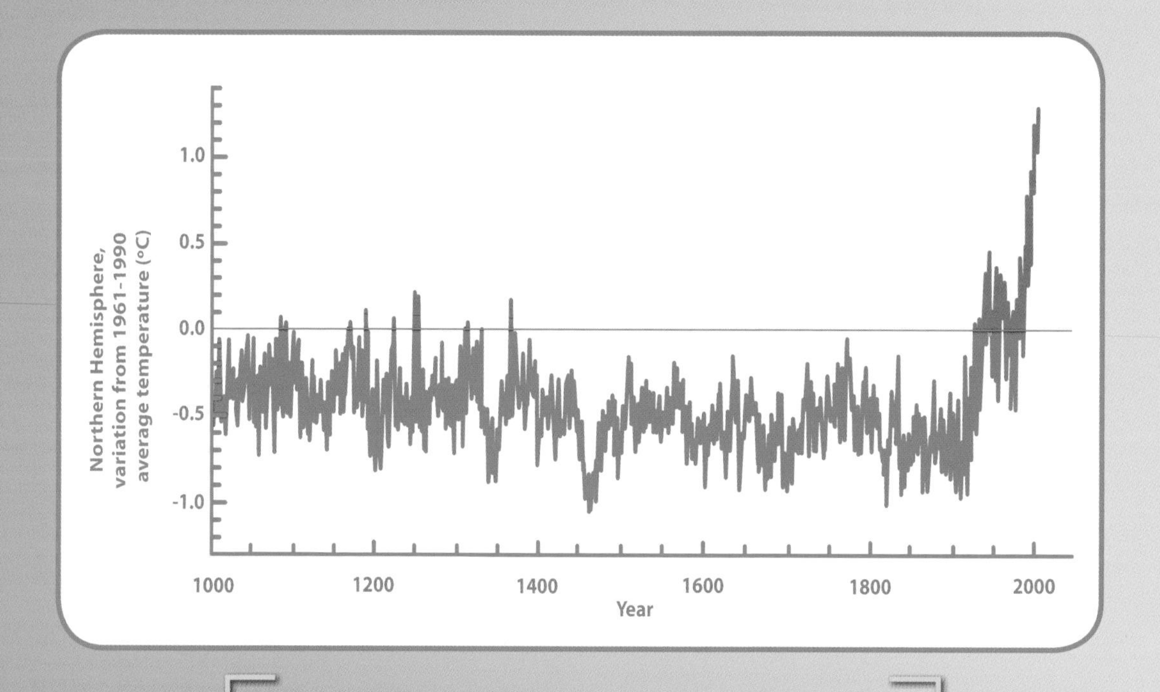

Nineteen of the warmest years on record (since thermometers were first used in the 1850s) have been in the last 20 years!

Melting Away

The Arctic is really feeling the heat. In the last 100 years, it has warmed twice as fast as the global average temperature. Why? Because much of the Arctic, the area surrounding the North Pole, is an ocean covered by ice. As the climate warms, more ice melts—especially during the summer. So more summer sunlight is absorbed by the water instead of being reflected by ice. That, in turn, heats the air even more.

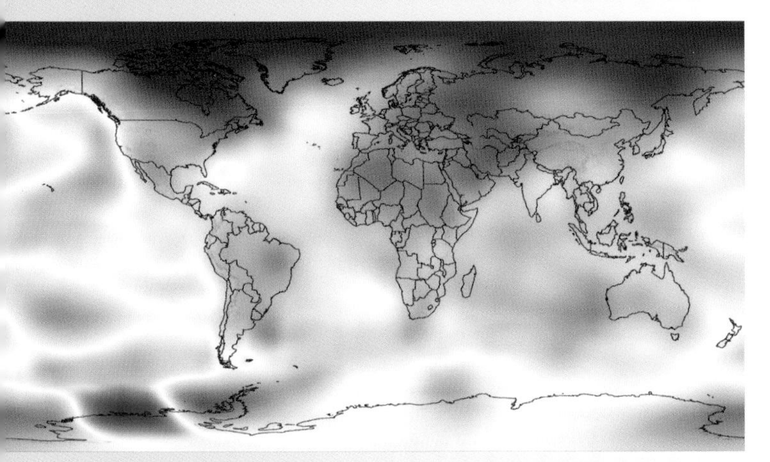

Change in surface temperature from 1880-2012 average (°C)

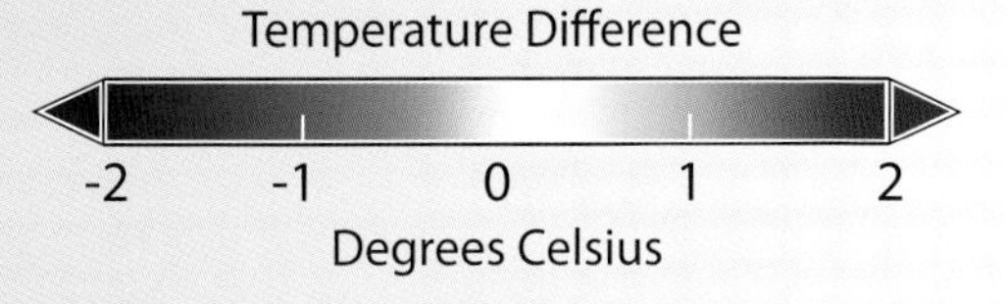

That doesn't mean that every spot on Earth is 0.8°C (1.5°F) warmer than it was 100 years ago. Some parts of the world have warmed more than others—land more than oceans, the northern parts of the planet more than the equator. Wow, look at the bright red Arctic.

Warming Signs

So it's getting warmer. What's the big deal? Well, scientists have already measured many real changes all around the planet. The oceans are warmer. Glaciers on mountains and ice caps are shriveling. There's more rain in the northeastern U.S., and storms are more intense. There's less rain in the parched southwestern U.S. Animals are scrambling to find new homes because they're no longer suited to the climate or to the changing ecosystem in their current ones.

A Walk on the Beach

Temperatures over land have been rising about twice as fast as over oceans. Why? Just walk barefoot on the beach on a hot day. When sunlight hits the beach, it's absorbed by the top layer of sand. The result? Sizzling feet! But when sunlight hits the water, it penetrates farther down than it could in the sand. Ripples and waves also mix the water. So the Sun's heat is shared with water farther down. The surface of the sand is much hotter than the surface of the water, so the sand radiates more heat back into the air.

HIGHWAYS IN THE SKY

Why does the warmer air have such widespread effects? It touches everything—the atmosphere ties the parts of our planet together. The air isn't just sitting there. It's in constant motion. As the air moves around the planet, it carries all sorts of really important stuff—like heat, water, and chemicals—with it. It crosses borders between countries without ever showing a passport.

Being There

The Eyes Have It

The world's fastest winds roar 6,096 meters (20,000 feet) above us in a twisting, rushing river of air called the jet stream, where wind speeds can reach 483 kilometers (300 miles) an hour. The jet stream is thousands of kilometers (miles) long, but only a few hundred kilometers (miles) wide. Next time you get a speck in your eye, remember this—that speck might have just blown in from the mountains of China or the deserts of Africa.

It's a Breeze!

What's the scientific term for air in motion? Hold on to your hats—scientists call it "wind." Wind starts whenever and wherever air in a particular place is heated up. This warmed air rises up. Cooler, heavier air nearby rushes into the space where the warmer air used to be. Whooosh—wind.

See You Later, Equator

The Sun's rays don't hit all parts of the planet with the same intensity. Earth's broad midsection, around the equator, receives more of the Sun's heat. This keeps the tropics toasty all year long. But because warm air rises, a lot of this heat hitches a ride on the winds. The atmosphere is *always* moving heat around the planet. Imagine vast highways in the sky where the traffic never stops—and all the trucks are carrying heat, heat, heat. Good thing. If this heat wasn't redistributed, the tropics would be unbearably hot, and the high latitudes would be a solid block of ice!

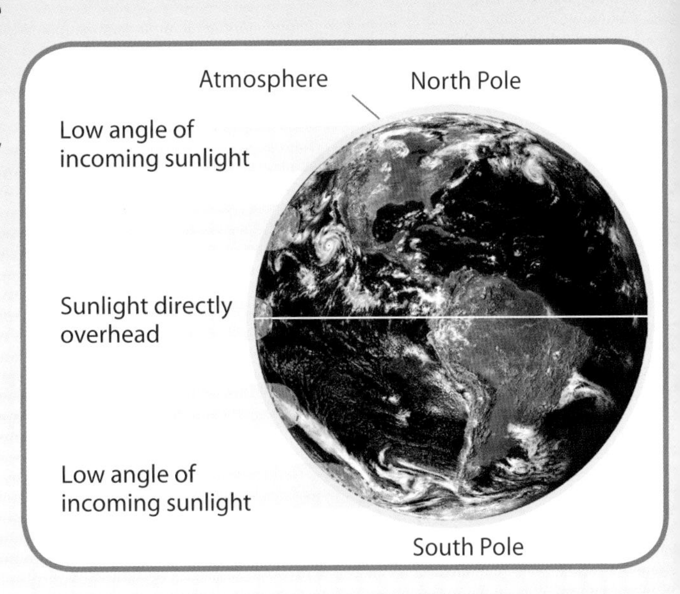

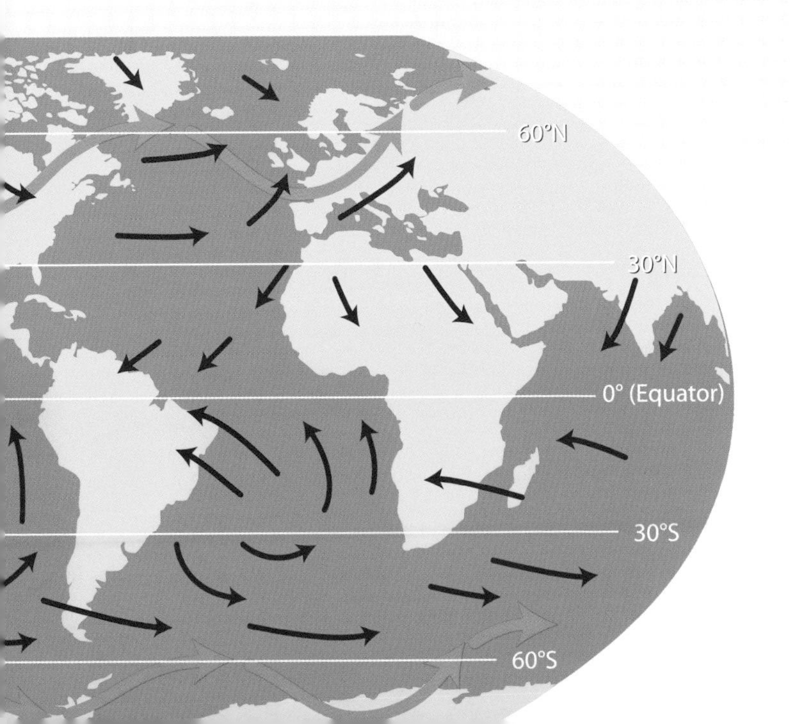

CO_2 ON THE MOVE

People are also affecting another of Earth's cycles—the carbon cycle. Do we care? *Yes*. The carbon cycle (below) continuously transports carbon around the planet, and it controls the amount of carbon dioxide in our atmosphere. You don't want to mess with *this* cycle!

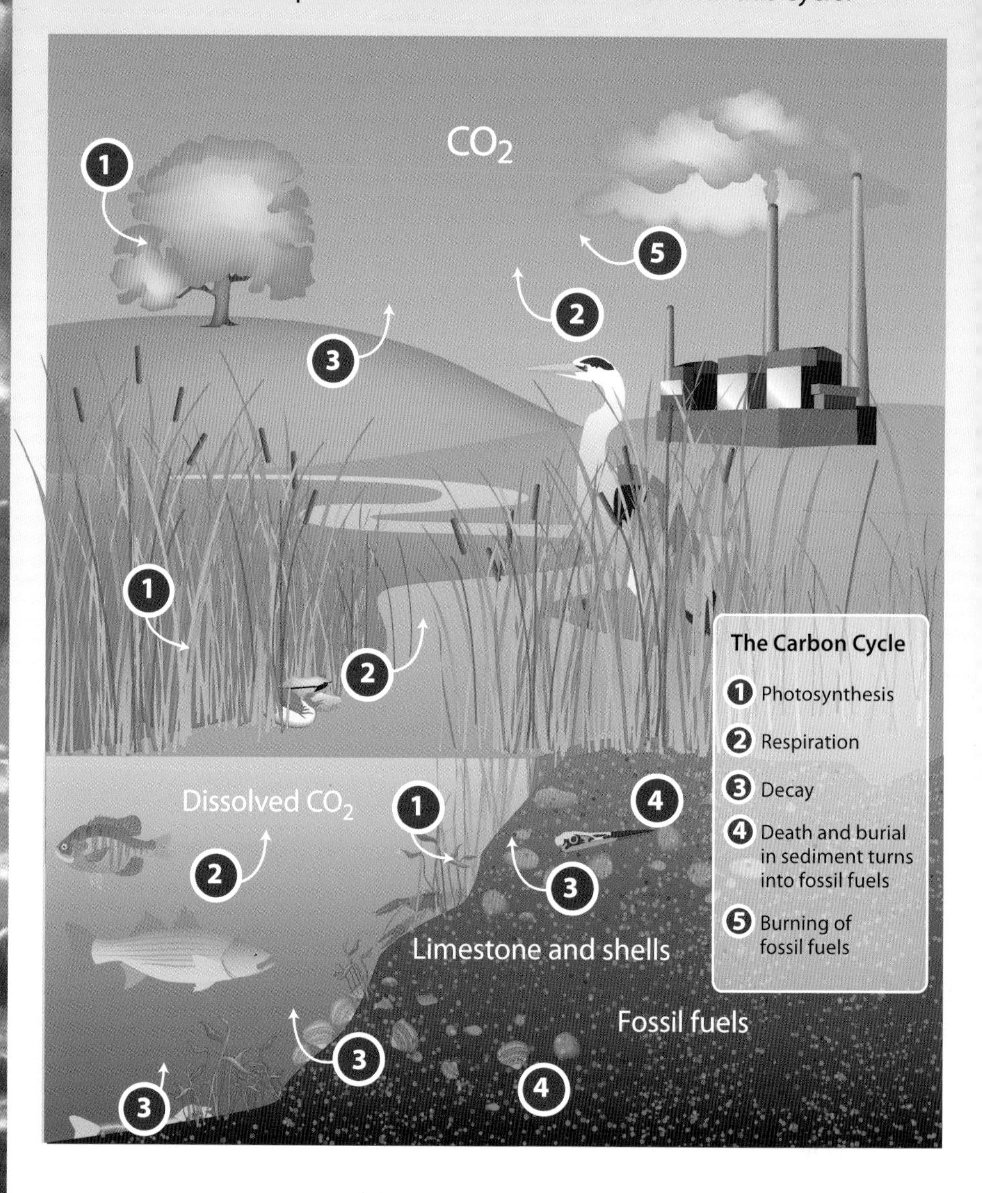

Carbon-a-Go-Go

There's only so much carbon on Earth. Some of it is deep underground in coal mines, some is in the lead of your pencil, some is in the shells of ocean animals, a lot is in plants, and some—a small amount—is in the form of carbon dioxide in the air. Over days, months, years, or eons, carbon shifts shape and moves between the oceans, the land, living things, and the air.

What's the Big Idea?

Where Has All the Carbon Gone?

For every ton of carbon in the atmosphere, there is about:

- 1 ton in biomass (trees and plants).
- 2 tons in the soil.
- 5 tons in fossil fuel deposits.
- 50 tons in the oceans.
- 35,000 tons in rocks.

Thank a Plant

For more than three billion years, certain organisms—first just microscopic phytoplankton in the ocean, now everything from floating algae to forests of redwoods—have been using energy from sunlight to convert carbon dioxide and water into food. We care because this is what keeps our air filled with oxygen. But we also care for another reason. While those organisms are pumping oxygen into the air, they're sucking carbon dioxide out of it.

What Goes Around . . .

Comes around. When plants die and then decay, much of the carbon dioxide that they took in and stored in their shoots, roots, leaves, and so on is returned to the atmosphere.

Hitch a Ride

The oceans are huge storage tanks. Carbon dioxide dissolves in cool seawater. Then it hitches a ride on ocean currents that act like massive conveyor belts, and heads down to the depths on a trip that lasts a thousand years. Eventually, it rises up and reaches warm surface waters near the equator and bubbles back out into the air. Carbon dioxide escapes from the ocean just as it does from a carbonated soda on a hot day.

A Shell Game

Some of the carbon dioxide dissolved in the oceans ends up being used for the undersea world's most popular building material—calcium carbonate. Corals use it to construct reefs. Sea urchins, clams, mussels, and countless other marine creatures build their shells out of it. These construction workers have built-in hard hats!

Burial at Sea

The carbon-rich shells and skeletons of ocean animals and plants eventually settle on the seafloor. Over millions of years they are pushed deeper and deeper down into the Earth. This carbon is *long* gone: It won't see the light of day again for millions of years—until it is released as gas from volcanoes or deep sea vents. Or when it is drilled or mined millions of years later as oil, coal, or natural gas.

Astronauts snapped this picture of an Alaskan volcano spewing ash and gas.

Short Circuit

When we extract coal or oil from deep beneath the ground to fuel our energy needs, we are bypassing the age-old natural pathways that return carbon to the atmosphere. Normally, it would remain trapped deep within the Earth. But when we burn these fuels to power our cars or computers, we are delivering it straight into the air—millions of years ahead of schedule.

Chop Chop

Trees, like all plants, take carbon dioxide out of the air. But for the last few centuries, we have been cutting them down all over the world. Since 1600, 90 percent of the forests that once grew in the continental U.S. have been cut down. In just the last four decades, nearly one-fifth of the Brazilian Amazon rainforest has been cleared. When trees are burned for fuel or to clear the land for agriculture, carbon dioxide is released. Deforestation is a double whammy, because it also means there are fewer trees around soaking up carbon dioxide for photosynthesis.

A swath of rainforest was cleared to make this road. Now people will use the road to cut down even more trees.

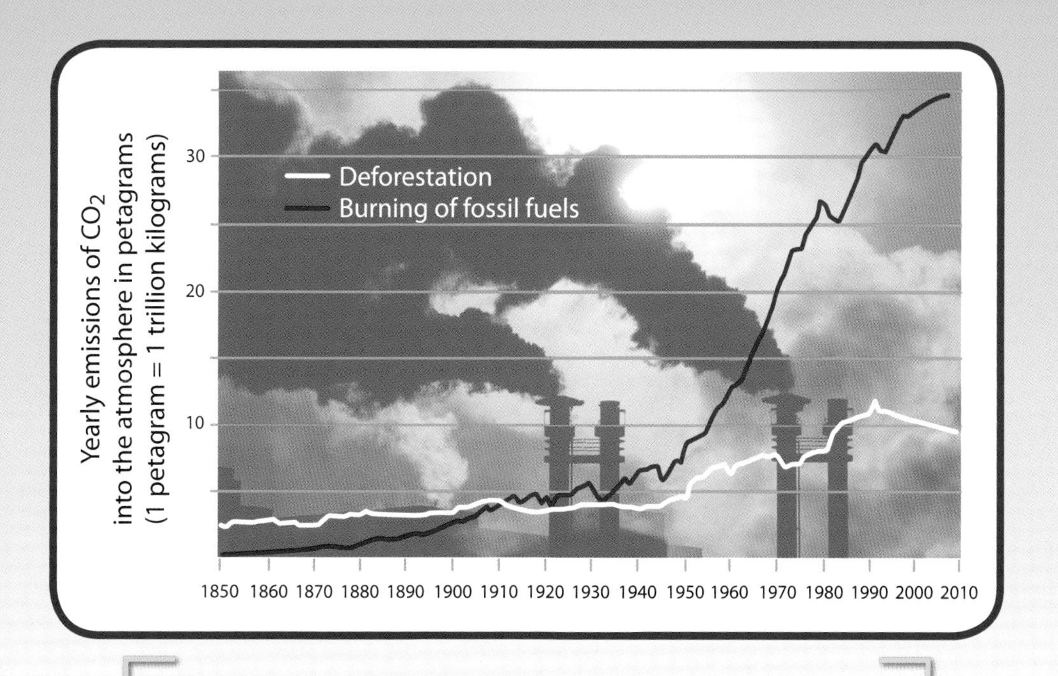

This graph shows the two big sources of carbon dioxide emissions into the air since 1850. It's easy to see which one has gone up the most!

Don't Hold Your Breath

Here's a carbon dioxide contribution you make every minute of every day, but you can't do much about it—breathing. Every time you exhale, you are releasing a little carbon dioxide into the atmosphere.

CLIMATE COMPLICATIONS

Have you ever spent a long day on the beach? At noon, you might sit under an umbrella for some cool shade. But then as the Sun sets, you might wrap yourself in a beach blanket to stay warm. Like giant umbrellas and blankets, clouds can also cool or warm Earth. Clouds work in both ways. They can cool Earth by reflecting the Sun's energy and by shading us. They also can warm Earth by trapping heat that rises from the ground. Clouds can affect the climate—but the climate can affect clouds, too. That makes them hard to understand—and really important!

Experts Tell Us Karen Shell

Atmospheric Scientist
Oregon State University

Earth's climate adjusts to change. But exactly how does it react to specific changes? To better understand this, scientists use the world's most powerful computers to study climates past, present, and future. In those virtual worlds, they toy with everything from plants on the ground to clouds in the sky. "It's like playing SimEarth all day. You could replace all the plants with dirt and see what happens," atmospheric scientist Karen Shell says. Karen uses computers to study how dust in the atmosphere can soak up or reflect the Sun's energy. Even little specks of dust kicked up by the wind in a drying world can make a big difference. Focusing on the very small has been a part of Karen's love for science ever since she grew bacteria in dishes and looked at them through a microscope when she was little. Now with computer models, she can see what the future might look like—without having to wait around.

Thin Blankets . . .

The clouds that warm Earth the most are high, wispy cirrus clouds. They're cold and don't cast much of a shadow—sunlight passes right through them. But when that sunlight is absorbed by the land, and heat is radiated back toward space, cirrus clouds do a good job of trapping it. With lots of cirrus clouds around the planet, Earth's temperature can rise.

In Latin, *cirrus* means "curl of hair" . . .

. . . and Thick Umbrellas

Down closer to the ground you'll find thick, fluffy cumulus clouds. They help keep things cool by reflecting sunlight back into space. They're sort of like shiny umbrellas.

. . . and *cumulus* means "heap."

Balancing Act

Clouds don't just tote water around the world. They also play a huge role in both warming and cooling Earth. In today's world, the warming effect of some clouds is more than canceled out by the cooling effect of . . . other clouds. In fact, the world would be a slightly *warmer* place if there weren't any clouds in the sky.

Being There

The Other Side of Clouds

Lying on the grass watching clouds isn't being lazy—you're conducting scientific observations. Knowing how many and what kinds of clouds are in the sky is important to understanding our climate. But it's impossible to spread cloud watchers all around the world—would *you* want to lie on your back in Antarctica? That's why scientists use wide-eyed satellites (right) to study clouds from above. Satellites can count clouds across the whole globe, even over remote deserts or oceans. They can also measure cloud size, altitude, temperature, and thickness. Nothing lazy about satellites, either.

Cloudy Future

Earth's climate will change as the world grows warmer. But how? That's a mystery partly shrouded in—you guessed it—clouds. Will a warmer world have more clouds in the sky—or fewer? And what types of clouds will they be? Will most be clouds that warm or clouds that cool? That's important, since even small changes in our planet's cloud cover can have big effects on climate.

Experts Tell Us — Richard Somerville

Theoretical Meteorologist
Scripps Institution of Oceanography

Have you ever gazed up at clouds and marveled at how quickly they change shape? Then you already understand something about what theoretical meteorologist Richard Somerville does. Richard studies clouds—but he looks at far more than he could see by himself. He uses information gathered from satellites, airplanes, and weather balloons to study what effect clouds will have on climate change. We know the world will be warmer, but by how much? That's about as predictable as the next shape a cloud will take. "A large chunk of the uncertainty is because we don't know how the clouds will react as the climate changes," he says. Richard became interested in the climate when he was a young boy. "I built all these weather instruments. I thought studying the weather was the coolest thing," he says. And now it's one of the hottest topics in science.

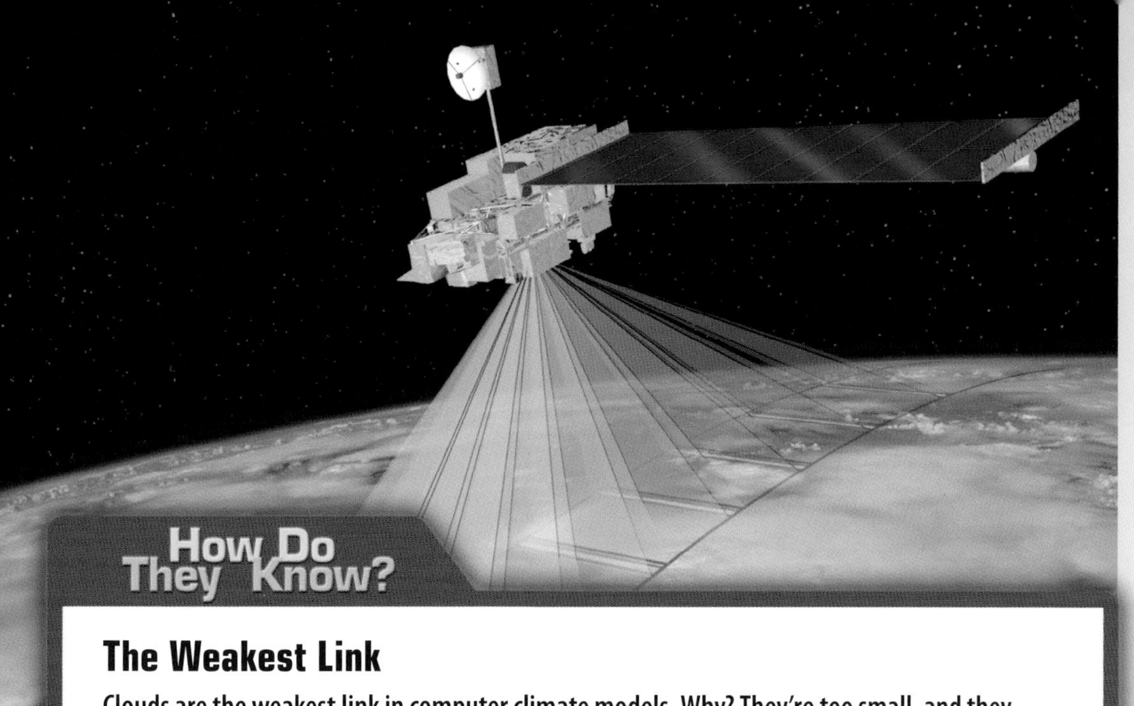

How Do They Know?

The Weakest Link

Clouds are the weakest link in computer climate models. Why? They're too small, and they change too quickly. Computer models pretend the atmosphere is divided into a grid of boxes. The problem is that with today's computers, each box is much bigger than most clouds. When scientists run the computer models, the grids just can't accurately predict how clouds act or how they contribute to climate change. But bigger and faster computers will eventually let scientists make the grid boxes smaller and the models more realistic. Sounds like they're talking about video games!

Smoke and Mirrors

Want another complication? The air is full of tiny particles of dust, dirt, and smoke called aerosols. Yes, they have an effect on our climate, too. Most are natural and come from dust shot into the sky from volcanoes or smoke billowing from forest fires. But others pour out of smokestacks and tailpipes. How do aerosols affect climate? They act like miniature reflectors—sunlight bounces off them and back into space. That cools Earth. Could that be a good thing?

The Gritty Truth

Aerosols also change clouds and, with them, rainfall patterns. Each droplet of water in a cloud has as its seed an aerosol particle. The water vapor condenses on the aerosol. If you put more of these gritty particles into the atmosphere, more—but smaller—droplets will form. Since smaller droplets are slower to fall from the sky, more aerosols could mean less rain.

CHANGE THE AIR, CHANGE THE PLANET

Call it season creep. As Earth's air warms, winters are getting shorter, spring is coming sooner, fall is starting later. Strange things are happening around us. The changes are being felt in the wild, where animals depend on the rhythms of nature for their survival. The changes are also being felt in suburbia, where spring is being sprung on backyard gardeners earlier than ever. Some day you may still be wearing summer clothes in late fall. Creepy.

Honk if You're Confused

Canadian geese are heading south earlier. Tree swallows are laying their eggs sooner. Frogs are mating earlier. Winter athletes in the eastern U.S. are losing ice time as frozen lakes thaw a full week sooner. Even breakfast is being disrupted—the maple syrup harvest is beginning earlier in the year, and pancakes are beginning to suffer.

Experts Tell Us: Claudia Tebaldi

Project Scientist
National Center for Atmospheric Research

Scientists say climate is what you expect and weather is what you get. So what do we get with global warming? "More heat waves and other extreme weather," says Claudia Tebaldi. Claudia helps make sense of different computer models of the climate and what they say about future weather. "They all project warmer temperatures and more precipitation," Claudia says. That means longer, more intense, and more frequent heat waves. We can crank up the air-conditioning, but for species that can't beat the heat, the changes are bad news. There also will be more rain—but fewer rainy days. Isn't that good news? Claudia reminds us that lots of rain all at once leads to floods. Even though Claudia looks decades into the future, she sees signs today of a warming, changing world everywhere around her.

It's Raining, It's Pouring

As the air warms, more water evaporates from the oceans. And what goes up . . . will eventually come back down—in this case, as rain or snow. Rainstorms are already becoming more intense. That's not good news in places that already get lots of rain. And even in deserts it's not necessarily good news. Too much rain all at once can lead to flash floods and be just as bad as too little rain.

Not-so-Perma Frost

Permafrost? What's that? It's ground that's been frozen since at least the last ice age, and it covers about a quarter of the Northern Hemisphere—part of the U.S., Canada, Europe, Russia, and a few other places. But as the air warms, the permafrost is starting to melt. As it melts, the icy ground turns to soggy ground. Already in Alaska roads are buckling and trees are tilting.

Compare the two photographs of Tanana Flats, Alaska. Over the past 20 years, the permafrost has melted, turning the tundra into wetlands.

Squish, Squish

Squishy ground is bad enough—but there's another problem. As the ground thaws, it sends more—potentially lots more—greenhouse gases into the air. As the bits of frozen grass and leaves in the permafrost soil thaw and decay, microbes release methane or carbon dioxide into the atmosphere. The amount of carbon in permafrost is *not* small. If all the permafrost in Russia thawed, the carbon dioxide released would double the amount in the atmosphere today. Yikes.

It was easy for astronauts to see the "eye" of Hurricane Isabel in 2003.

Stormy Weather

Global warming will trigger more extreme and more unpredictable weather. In Florida and along the Gulf Coast, hurricane season already is a dangerous time. People who live on the coast often have to board up their windows and move to higher ground. More and possibly bigger storms could do even more damage—not only to homes, but also to the rich ecosystems in coastal waters.

Hurricane Chasers

When everyone else at MacDill Air Force Base in Florida is locking things down in anticipation of a big storm, "Miss Piggy" is getting ready to take off. This is one of only two airplanes built to fly into the eye of a hurricane to study how hurricanes form, develop, and die out. Pilots fly into the center of the hurricane and drop instruments into the storm that tell weather forecasters if it's getting stronger or weaker.

This fishing pier used to be over the lake!

Bone Dry

Lake Mead near Las Vegas, Nevada, is a natural yardstick for what's going on in the wider Colorado River Basin. And what's going on is an extended drought. The lake is 27 meters (87 feet) below its high-water mark. Lake levels always fluctuate. But in a warmer world, there will be more—and more intense—droughts.

Can you find the islands that rejoined the shoreline when Lake Mead shrank between 2000 and 2003?

Heat Sink

The air is already warmer—but we'd all be much toastier if it weren't for the oceans. They have been sponging up heat and bailing us out big-time. A lot of extra heat has resulted from our greenhouse gas emissions—and the oceans have absorbed 20 times more of it than the atmosphere. If all the extra heat that the oceans are storing were suddenly released, the air temperature would rise about 22°C (40°F). A pleasant 27°C (81°F) day would become an unbearable 49°C (120°F) in the shade. Then we'd really be in hot water.

The Future Atmosphere

Look up. The air seems to go on forever. But if you look down on it from space, you see that our atmosphere is really, really thin.

Even 50 years ago, it would have been hard to believe that people could actually change the atmosphere of a planet. But that's exactly what's happening. The good news is, there's no reason we can't change it back.

Glossary

atmosphere (n.) the mixture of gases (nitrogen, oxygen, and traces of others) surrounding Earth, held in place by the force of gravity (pp. 4, 6, 7, 8, 9, 12, 13, 14, 15)

carbon cycle (n.) the never-ending movement of carbon between the atmosphere, oceans, land, and living organisms (p. 26)

climate (n.) prevailing weather conditions for an ecosystem, including temperature, humidity, wind speed, cloud cover, and rainfall (pp. 10, 11, 18, 22)

fossil fuel (n.) a nonrenewable energy resource such as coal, oil, or natural gas that is formed from the compression of plant and animal remains over hundreds of millions of years (pp. 15, 26, 29)

greenhouse effect (n.) the warming that occurs when certain gases (greenhouse gases) are present in a planet's atmosphere. Visible light from the Sun penetrates the atmosphere of a planet and heats the ground. The warmed ground then radiates infrared radiation—heat—back toward space. If greenhouse gases are present, they absorb some of that infrared radiation, trapping it and making the planet warmer than it otherwise would be. (pp. 12, 13)

greenhouse gas (n.) a gas such as carbon dioxide, water vapor, or methane that absorbs infrared radiation, or heat. When these gases are present in a planet's atmosphere, they absorb some of the heat trying to escape the planet instead of letting it pass through the atmosphere. The resulting warming is called the greenhouse effect. (pp. 12, 13, 14, 15, 19, 36)

photosynthesis (n.) the process by which plants and other photosynthetic organisms use energy from sunlight to build sugar from carbon dioxide and water. As part of this process, oxygen is released. (pp. 9, 17, 29)

Answers

4 U 2 Do, page 9
Your "nitrogen" pile should have 78 pieces. Your "oxygen" pile should have 21 pieces. Your "trace gases" pile should have 1 piece.

4 U 2 Do, page 21
Some water in the jar without the lid has evaporated. In the jar with the lid, the droplets are from water evaporating and then condensing.

Index